Our Solar System
The Sun and Its Effects

by Glen Phelan

Table of Contents

Develop Language 2

CHAPTER 1   The Seasons and the Sun 4
                     Your Turn: Predict 9
CHAPTER 2   Objects in Our Solar System 10
                     Your Turn: Classify 15
CHAPTER 3   A Closer Look at the Sun 16
                     Your Turn: Summarize 19

Career Explorations 20
Use Language to Show Sequence 21
Science Around You 22
Key Words 23
Index 24

WASHINGTON SCHOOL
Millmark
EDUCATION

DEVELOP LANGUAGE

Earth has four seasons during the year: winter, spring, summer, and fall. Look at the photos and tell what they show about each season.

Think about what each season is like where you live. Then answer these questions:

What is the weather like in winter?

How can plants change in the spring?

What is the weather like in summer?

What can happen to trees in fall?

What season is it now where you live? How can you tell?

winter snow

Our Solar System: The Sun and Its Effects

spring flowers

fall leaves

summer heat

Develop Language 3

CHAPTER 1

The Seasons and the Sun

Many people spend more time outside in summer than in winter. There is more daylight in summer, and the weather is often warmer.

During fall, the weather is cooler and there are fewer hours of daylight. Winter has the coldest weather and the fewest hours of daylight. In some places, snow falls.

By spring, there are more hours of daylight and the weather is warmer. Summer follows spring and the **cycle** of the seasons continues.

Why does the weather and amount of daylight change each season? The answer has to do with Earth's position and motion in space.

cycle – a thing that happens again and again in the same order

Rotation

Earth spins, or **rotates**, around an imaginary line called an **axis**. Earth's axis is tilted.

Because Earth is shaped like a ball, half of it is always facing the sun. On this half, it is daytime. The half that faces away from the sun is in darkness, and it is nighttime.

You can see in the picture that Earth rotates from west to east. This rotation makes the sun look like it rises in the east, moves across the sky, and sets in the west.

But the sun is not moving around Earth. Earth is rotating on its axis.

rotates – spins around an axis
axis – an imaginary line around which an object

Explore Language
COMPOUND WORDS
day + time = daytime
night + time = nighttime

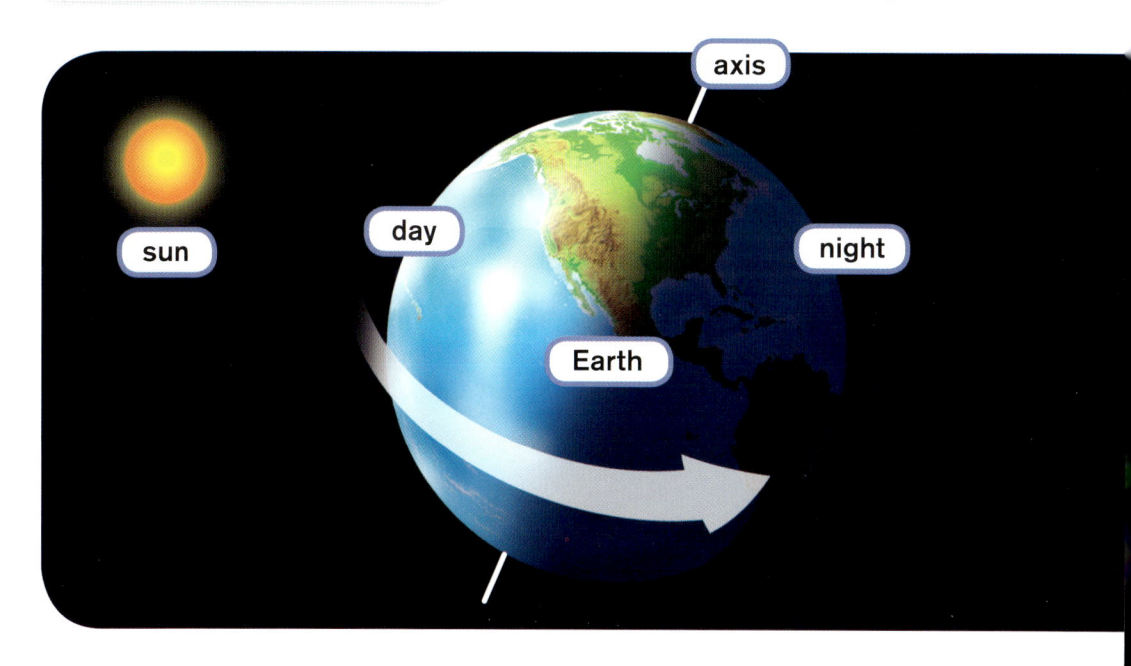

▼ Earth's rotation causes day and night.

Chapter 1: The Seasons and the Sun 5

Revolution

In addition to rotating on its axis, Earth also **revolves** around the sun. This motion is called revolution. Earth makes one revolution around the sun in one year. As Earth revolves, the seasons change.

Notice that Earth's axis is tilted the same way no matter where Earth is in its **orbit**, or curved path, around the sun.

revolves – moves around something
orbit – the path of an object around another object

Because of this tilt, different parts of Earth receive different amounts of sunlight throughout the year.

▼ **Here, it is spring in the northern part of Earth. Now both halves of Earth are receiving about the same amount of direct sunlight.**

spring

▶ **Here, the northern part of Earth is tilted toward the sun. It is now summer in the northern part, because it is receiving the most direct sunlight.**

summer

6 *Our Solar System: The Sun and Its Effects*

The diagram shows Earth's position in space at the beginning of each season for the northern part of Earth. Follow Earth's orbit in the diagram as you read about each season.

SHARE IDEAS Look at the diagram. **Explain** why the southern part of Earth has summer when the northern part has winter.

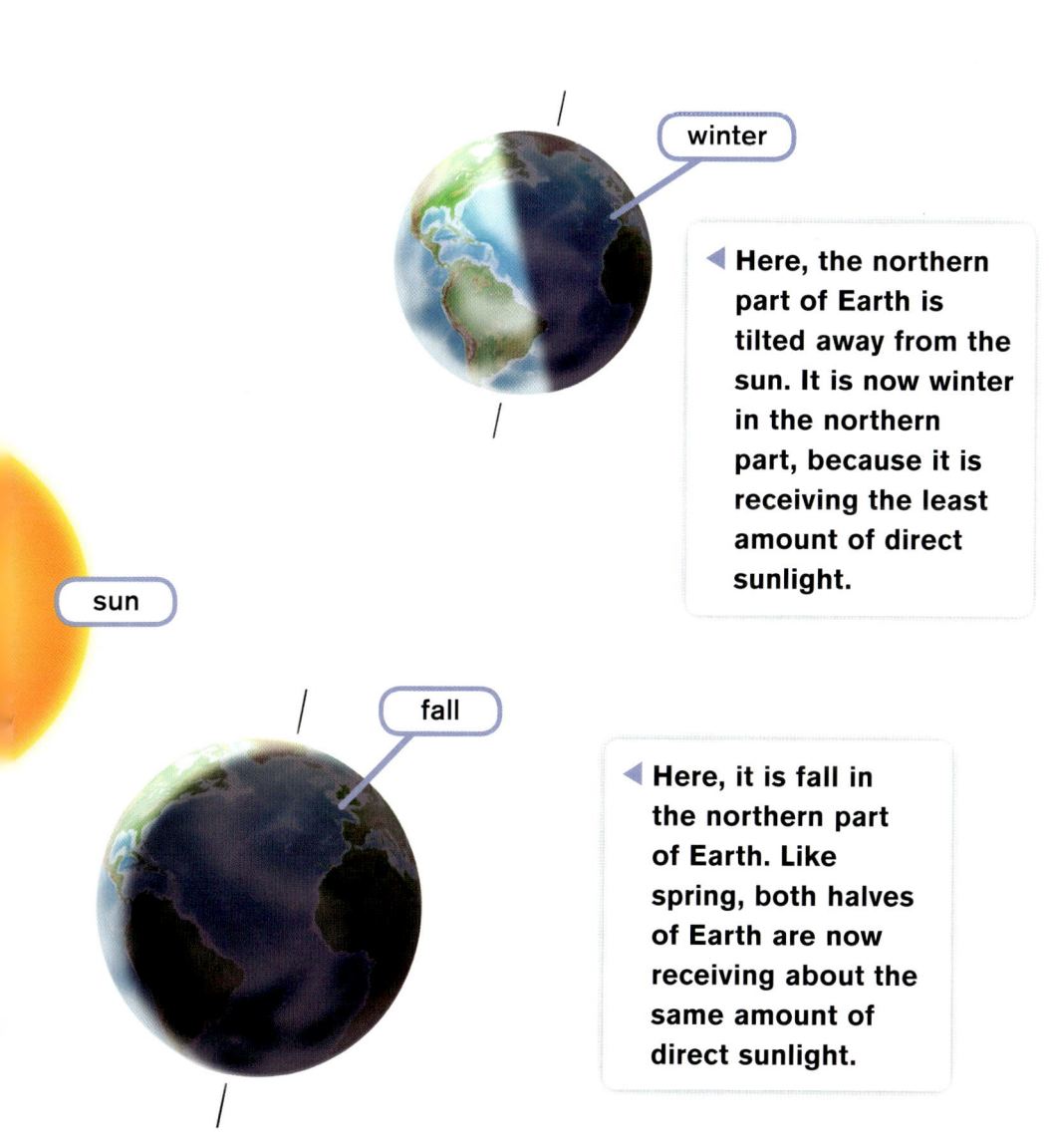

winter

◀ Here, the northern part of Earth is tilted away from the sun. It is now winter in the northern part, because it is receiving the least amount of direct sunlight.

sun

fall

◀ Here, it is fall in the northern part of Earth. Like spring, both halves of Earth are now receiving about the same amount of direct sunlight.

Chapter 1: The Seasons and the Sun 7

Gravity Keeps Earth in Orbit

Year after year, Earth revolves around the sun. Why does Earth keep traveling in its orbit? The answer is **gravity**.

Gravity is a pulling force between two objects. The sun and Earth pull on each other. The sun's pull is larger, so its gravity keeps Earth in orbit.

How does pulling keep something moving around and around? Look at the girl. She pulls on the wire as she spins to keep the ball revolving around her. When she lets go of the wire, she stops pulling and the ball travels away from her.

Earth would travel away from the sun if the sun's gravity did not pull on it. But gravity keeps Earth in orbit and continues the cycle of the seasons.

gravity – a pulling force between two objects

◀ This girl is practicing for a hammer throw competition.

KEY IDEAS The rotation of Earth causes day and night. Earth has seasons because Earth is tilted as it revolves around the sun. The sun's gravity keeps Earth in orbit.

YOUR TURN

PREDICT

You know that Earth is tilted on its axis. But suppose Earth was not tilted. Suppose Earth's axis were straight-up-and-down, as shown in the drawing. What predictions could you make?

1. Predict how the number of daylight hours would be different on a straight-up-and-down Earth than those on a tilted Earth.

2. Predict whether or not a straight-up-and-down Earth would have seasons. Explain your prediction.

MAKE CONNECTIONS

How does the force of gravity affect your life? Think of ways gravity might have affected something you did today.

USE THE LANGUAGE OF SCIENCE

Why does Earth's tilt cause seasons as Earth revolves around the sun?

The tilt causes different parts of Earth to receive more or less direct sunlight during certain times of the year.

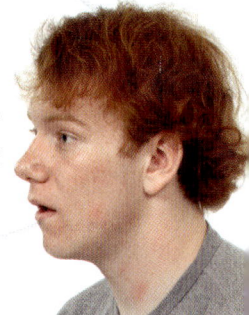

Chapter 1: The Seasons and the Sun

CHAPTER 2
Objects in Our Solar System

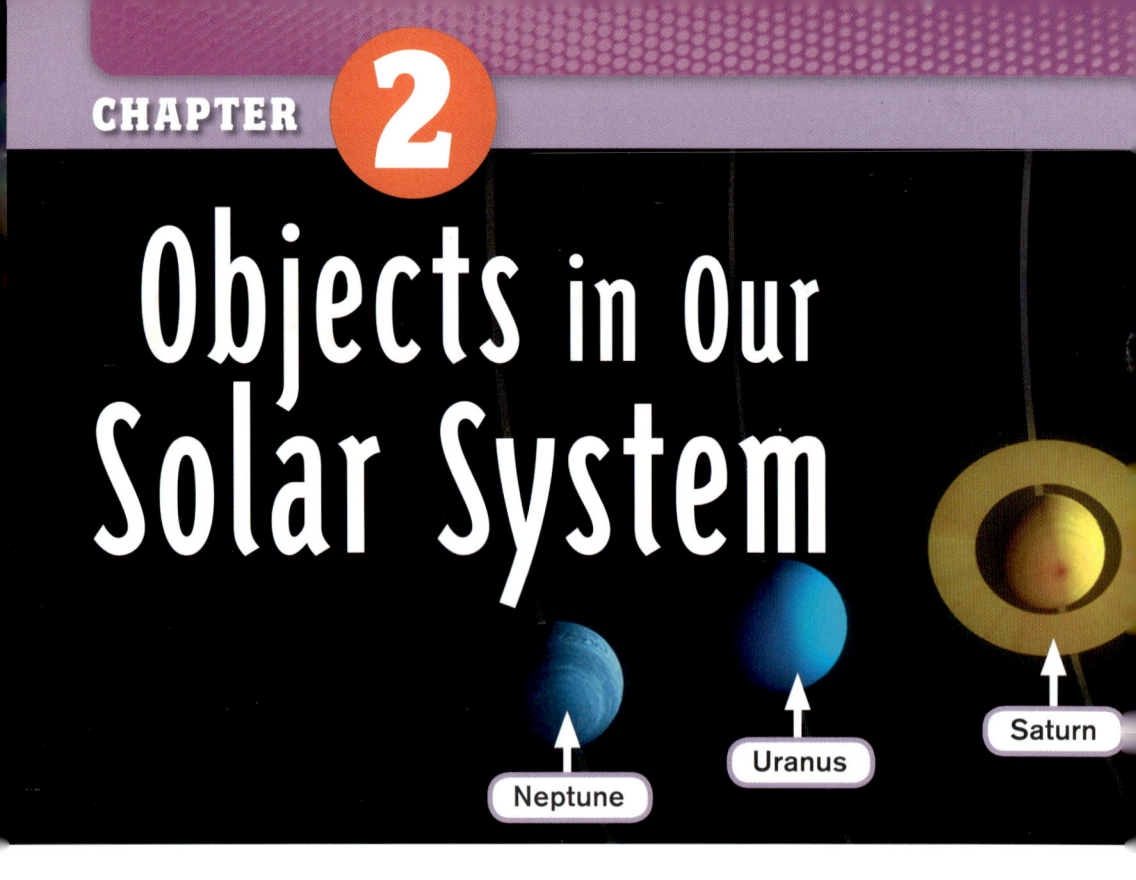

Neptune

Uranus

Saturn

The pull of the sun's gravity keeps Earth and millions of other objects in orbit around the sun. All of these objects, as well as the sun, make up our **solar system**.

Some of the largest objects in our solar system are the eight planets that are shown in the diagram above. Look at the names of the planets and their order from the sun.

Our solar system also includes moons. You probably already know that Earth has a moon, but did you know that other planets have moons, too? In fact, the planet Jupiter has more than 60 moons! A planet's gravity keeps each moon in orbit around the planet.

solar system – the sun and all the objects that revolve around the sun

Jupiter
Mars
Earth
Venus
Mercury

These observatories protect large telescopes.

Scientists have learned a lot about the planets and other objects in our solar system. One of the most important tools scientists use is a **telescope**, a tool that helps them see faraway objects.

Telescopes come in many shapes and sizes. Some telescopes are very large and have their own buildings, called observatories. Some telescopes travel in space.

The Hubble Space Telescope travels around Earth. It gives scientists very clear pictures of objects in space since it is above the atmosphere, which interferes with viewing.

telescope – a tool that allows people to better see faraway objects

Chapter 2: Objects in Our Solar System   11

Probes and Planets

Scientists also use **space probes** to learn about our solar system. These spacecraft carry cameras and other instruments, but no people.

Space probes send back pictures and other information for scientists to study. Much of what we know about planets comes from space probes.

▲ The space probe *Spirit* was built to explore Mars.

space probes – spacecrafts that carry instruments, but no people, to explore space

INNER PLANETS	
Planet	Features
Mercury	Mercury has many round craters that formed when rocks from space crashed into the surface. Nights on Mercury are freezing cold and days are hot enough to melt metal.
Venus	The thick clouds that cover Venus all the time trap heat on the planet and keep Venus very hot. The surface is mostly made from liquid rock that flowed up from beneath the surface, cooled, and hardened into solid rock.
Earth	Earth's water, air, and mild temperatures make life possible. It is the only planet known to have liquid water and living things.
Mars	The soil and rocks on Mars give the planet a red color. Mars has deep, wide canyons and tall mountains. It has huge ice caps but no known liquid water. It may have had liquid water in the past.

Our Solar System: The Sun and Its Effects

Scientists classify the planets into two groups: the inner planets and the outer planets. The inner planets are the four closest to the sun: Mercury, Venus, Earth, and Mars.

The inner planets are all small compared to the outer planets. They are all solid planets made of rock and are sometimes called the rocky planets.

The outer planets are the four farthest from the sun: Jupiter, Saturn, Uranus, and Neptune.

The outer planets all have a small, rocky center surrounded by liquid and a thick layer of colorful gases and clouds. They all have rings made of rocks, dust, and ice that circle the planet. Also, the outer planets are huge. Sometimes they are called the gas giants.

OUTER PLANETS

Planet	Features
Jupiter	Jupiter has a storm of rotating winds called the Great Red Spot. This storm has lasted for hundreds of years. Jupiter has over 60 moons.
Saturn	Saturn has the widest and most amazing rings of any planet. Saturn has at least 47 moons.
Uranus	The gases on Uranus give the planet a blue-green color. Uranus is the only planet that rotates on its side. It has at least 27 moons.
Neptune	A gas on Neptune gives the planet a blue color. Neptune also has storms, including one large storm called the Great Dark Spot. Neptune has 13 known moons.

Chapter 2: Objects in Our Solar System

Other Objects in Our Solar System

Scientists used to call Pluto the ninth planet. Then they discovered that this ball of rock and ice is not very different from many other objects in that part of space. Now Pluto is called a **dwarf planet.**

Pluto is not the only dwarf planet in the solar system. Between Mars and Jupiter is a ring of rocky objects called **asteroids**. Most asteroids are about the size of a small town. However, the largest asteroid, Ceres, is about the size of Texas and is large enough to be a dwarf planet.

Beyond Neptune and Pluto, there are millions of chunks of ice and rock called **comets** that orbit the sun. Some comets have orbits that bring them closer to the sun. As they get close to the sun, the ice on the comet turns to gas and forms a tail.

dwarf planet – an object in the solar system that is larger than most objects, but smaller than a planet

asteroids – chunks of rock that revolve around the sun

comets – chunks of ice and rock that revolve around the sun

KEY IDEAS Our solar system includes the sun and all the objects that revolve around it. We can explore our solar system using tools such as telescopes and space probes.

◀ **The Comet Hale-Bopp was visible in 1997. Here it is seen over observatories in Hawaii.**

Our Solar System: The Sun and Its Effects

Your Turn

CLASSIFY

The photos show three planets. Look at the photos carefully. Then read the descriptions in the chart. Finally, classify each planet as an inner planet or outer planet.

Planet	Inner or Outer
A. Has a storm called the Great Red Spot and is mostly made of colorful gases and clouds	
B. Has many craters on its rocky surface	
C. Its rocks and soil have a red color	

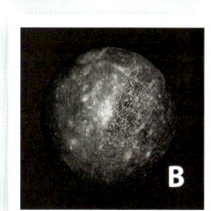

MAKE CONNECTIONS

The planet Uranus rotates on its side. Look on page 5 at how Earth rotates. Then make a drawing to show how Uranus rotates.

STRATEGY FOCUS

Monitor Comprehension

What ideas were difficult to understand in this chapter? Think about what you did to improve your understanding.

Tell when you used these strategies:
- rereading
- asking a question
- finding an answer in the text
- restating information

Chapter 2: Objects in Our Solar System   15

CHAPTER 3

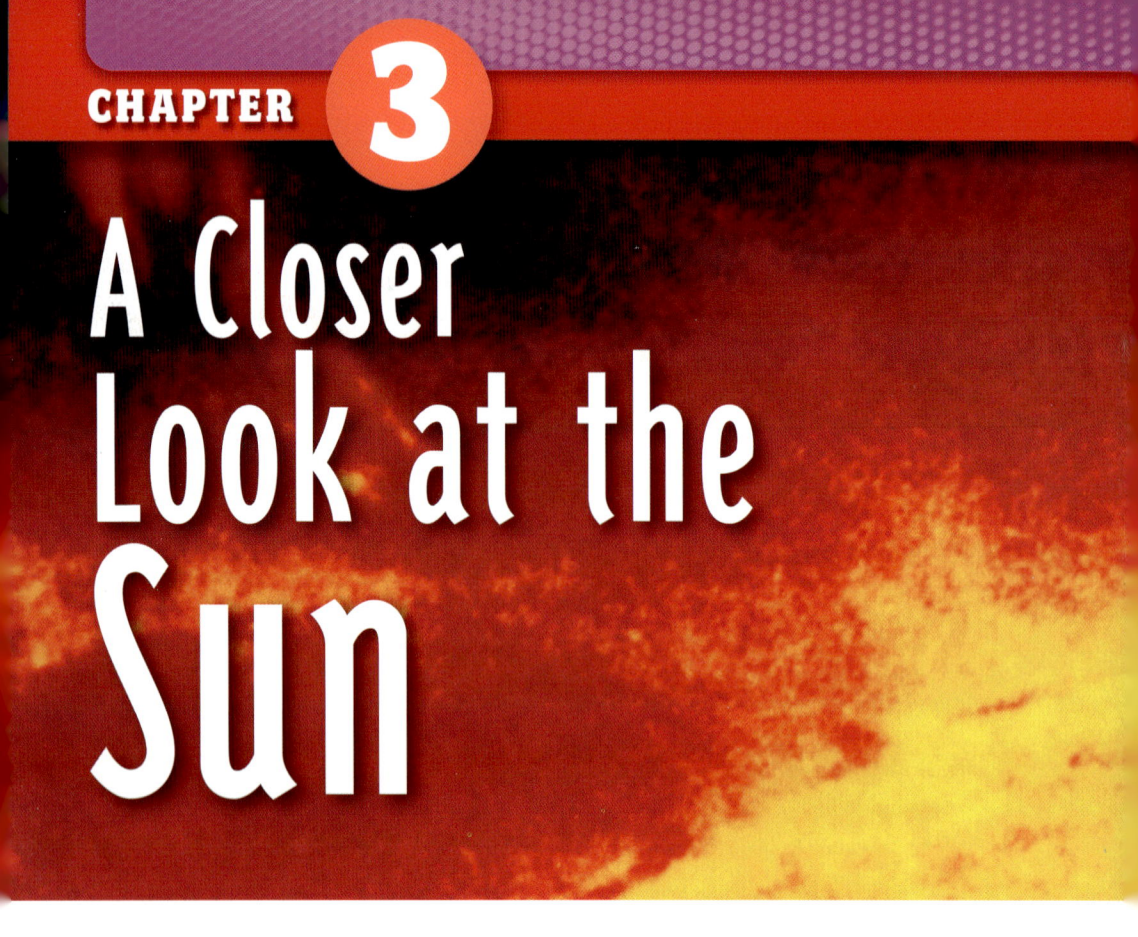

A Closer Look at the Sun

Can you name the brightest star in the sky? It's the sun! The sun is the center of our solar system.

In the night sky, stars look like pinpoints of light. Yet about half of them are larger than the sun. The sun looks much larger than any other star because it is much closer to us. Yet it is still 93 million miles away.

Like other stars, the sun is a glowing ball of hot gases. The sun's light and heat are produced when tiny particles of matter, called **atoms**, join together in the center of the sun. Some of the energy produced by the sun escapes out into space. Some of the sun's energy reaches Earth. It provides warmth and allows plants to grow.

atoms – tiny particles of matter

The sun is so large that one million Earths could fit inside it.

← solar flare

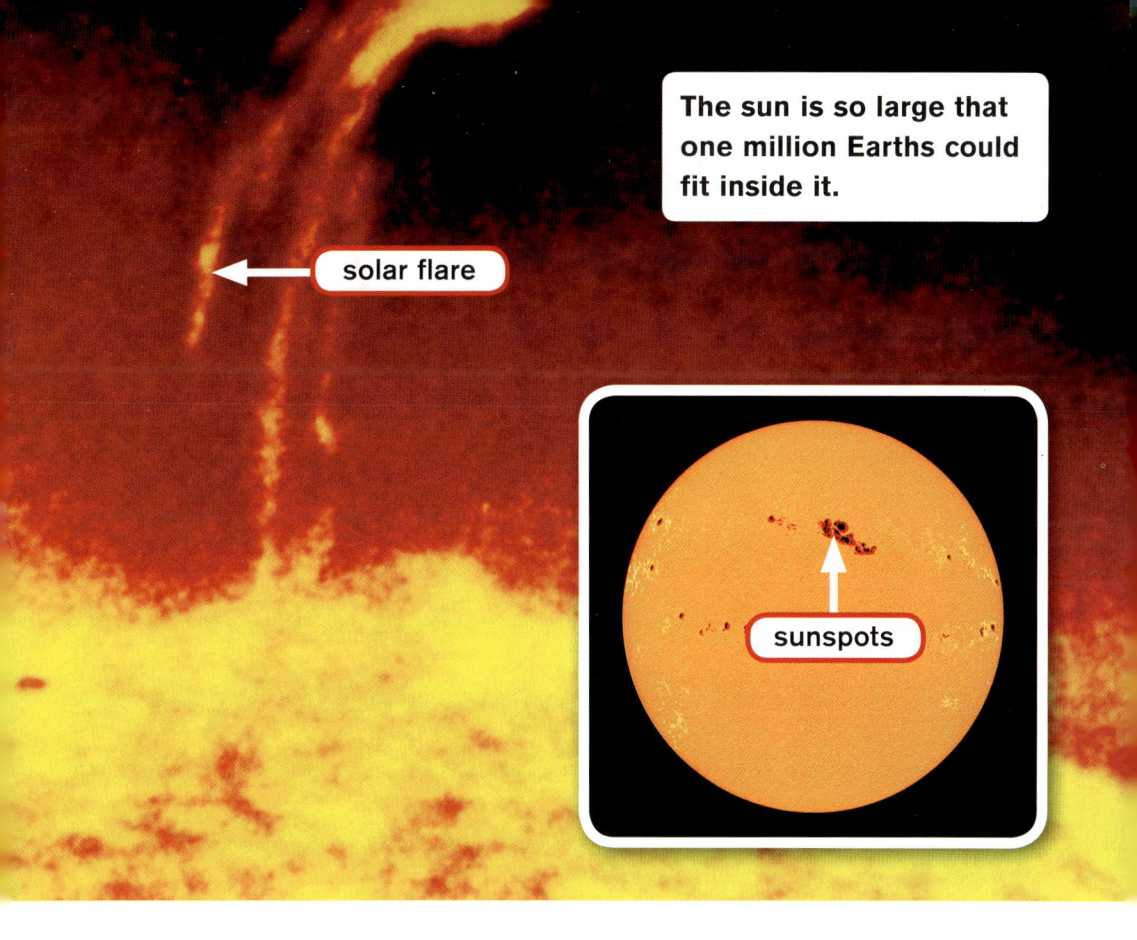

sunspots

Scientists see some interesting features when they study the sun, including **sunspots**. These are areas on the sun's surface where the gases are cooler than the gases around them. The cooler gases don't give off as much heat, so sunspots look dark.

Scientists also see **solar flares**, that occur above the sun's surface. Most solar flares occur around sunspots.

sunspots – areas on the sun's surface where the gases are cooler than the gases around them

solar flares – explosions above the sun's surface

Chapter 3: A Closer Look at the Sun

Effects of Solar Flares

When gas erupts from the sun during a solar flare, a burst of energy escapes into space. Some of this energy can reach Earth and disrupt radio signals. The energy can also cause power outages.

Solar flares can also cause colorful light displays in the night sky. These displays are called **auroras**. Usually, auroras can only be seen from places near Earth's poles, the most northern and southern points on Earth.

auroras – displays of light in the sky near the poles which are sometimes caused by solar flares

This aurora was seen over Denali National Park in Alaska.

KEY IDEAS The sun, like other stars, is a ball of hot gases that gives off energy. The sun's energy has effects on Earth.

18 *Our Solar System: The Sun and Its Effects*

YOUR TURN

SUMMARIZE

Review the headings, pictures, and text in this chapter. Then complete a chart like the one at the right by stating the main idea of each page.

Page	Main Idea
16	
17	
18	

MAKE CONNECTIONS

Think of all the ways the sun is important in your life. Make a list. Share your ideas with classmates.

EXPAND VOCABULARY

A compound word is a combination of two or more words. Compound words can be closed (spacecraft); open (space probe) or hyphenated (straight-up-and-down). Make a chart like this one. Look in the book to find examples of all three types of compounds. Draw pictures to illustrate some of the words you find.

Closed	Open	Hyphenated

Chapter 3: A Closer Look at the Sun

CAREER EXPLORATIONS

Science Journalists Inform the Public

Look through a newspaper or a newsmagazine. Watch the news on television or listen to a radio report. All of these stories were written by journalists.

Some journalists write stories about science. They interview scientists to find out the latest discoveries. Then they write articles to share the information with the public.

Below are some characteristics of the job. Think about which of these are positives and which are negatives for you. Then discuss whether you would like to be a journalist.

Journalists…

- talk with scientists and other newsmakers.
- do not spend all day in an office.
- write under the pressure of tight deadlines.
- write stories that millions of people read.
- have to respond to news events right away, even when they have made other plans.

▲ Chris Voohees, a mechanical systems engineer with the Mars Rover Exploration team, responds to a question during an interview.

USE LANGUAGE TO SEQUENCE

Words that Show Sequence

A sequence is a way to put things in a certain order. When you talk about a sequence of events, you tell about when certain things happen. You describe what happens first, next, and last. You may also describe what is happening while other things happen. You can show sequence with prepositions of time and adverbs of time.

EXAMPLE

Prepositions of time

During a year, Earth has four seasons.
Summer is the season **after** spring.

Adverbs of time

First there was a solar flare. **Then** energy escaped out into space. **Finally** an aurora formed.

- With a friend, look at the photos on pages 2-3. Describe the sequence of events in one photo. Use words such as **before**, **first**, **then**, **during**, and **finally**.

Write Using Sequence

Describe the different seasons in the place where you live. Tell what changes you see from season to season.

- Describe the weather in each season.
- Tell about each season in sequence.
- Use prepositions of time and adverbs of time.

Words You Can Use	
Prepositions of Time	**Adverbs of Time**
before during after	first then next finally

Use Language to Sequence

SCIENCE AROUND YOU

Rare Solar Eclipse Today

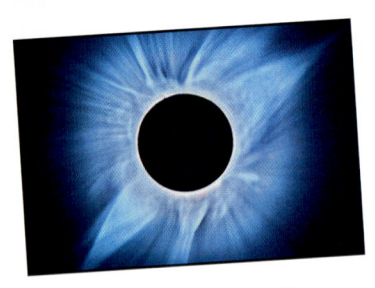

This afternoon we will witness an event that will not be seen in this area for another 350 years—a total solar eclipse.

Usually the moon passes a little above or below the sun in the sky. But at exactly 2:45 P.M. today, the moon will pass directly between Earth and the sun. For about 5 minutes, our town and the surrounding areas will be in the moon's shadow. The sky will be dark enough to see stars.

While the moon blocks the sun during those 5 minutes, a halo of light called the corona can be seen. This outer layer of the sun is too faint to see except during a total eclipse or with special telescopes.

Astronomers warn people not to look directly at the sun. People can see pictures on a local news station or on the Internet.

Read the newspaper article.

- What happens when the moon passes directly between Earth and the sun?

- What is the sun's corona?

Key Words

asteroid (asteroids) a chunk of rock that revolves around the sun
The **asteroid** is as big as a city.

axis (axes) an imaginary line around which an object spins
Earth's **axis** goes through the North and South Poles.

comet (comets) a chunk of ice and rock that revolves around the sun
Some **comets** come close enough to Earth to see without a telescope.

dwarf planet (dwarf planets) an object in the solar system that is larger than most objects, but smaller than a planet
Pluto is now called a **dwarf planet**.

gravity a pulling force between two objects
The sun's **gravity** keeps Earth in its orbit.

orbit (orbits) the path of an object around another object
Earth takes one year to complete its **orbit** around the sun.

revolve move around something
The moon **revolves** around Earth.

rotate spin around an axis
Earth **rotates** around its axis.

solar flare (solar flares) an explosion above the sun's surface
A **solar flare** often occurs near a sunspot.

solar system the sun and all the objects that revolve around the sun
The eight planets and their moons are part of our **solar system**.

space probe (space probes) a spacecraft that carries instruments, but no people, to explore space
Space probes send back close-up photographs of planets.

sunspot (sunspots) area on the sun's surface where the gases are cooler than the gases around it
A **sunspot** looks darker than the rest of the sun's surface.

telescope (telescopes) a tool that allows people to better see faraway objects
You can use a **telescope** to explore objects in space.

Key Words 23

Index

asteroid 14

axis 5–6, 9

aurora 18, 21

comet 14

cycle 4, 8

dwarf planet 14

eclipse 22

energy 16, 18, 21

gravity 8–10

moon 10, 13, 22

orbit 6–7, 8, 10, 14

planet 10-15

revolution 6

revolve 6, 8, 14

rotation 5, 8

rotate 5, 13, 15

seasons 2–4, 6–9, 21

solar flare 17–18, 21

sunspot 17

telescope 11, 14

MILLMARK EDUCATION CORPORATION
Ericka Markman, President and CEO; Karen Peratt, VP, Editorial Director; Lisa Bingen, VP, Marketing; David Willette, VP, Sales; Rachel L. Moir, VP, Operations and Production; Shelby Alinsky, Editor; Mary Ann Mortellaro, Science Editor; Amy Sarver, Series Editor; Betsy Carpenter, Editor; Guadalupe Lopez, Writer; Kris Hanneman and Pictures Unlimited, Photo Research

PROGRAM AUTHORS
Mary Hawley; Program Author, Instructional Design
Kate Boehm Jerome; Program Author, Science

BOOK DESIGN Steve Curtis Design

CONTENT REVIEWER
Tom Nolan, Operations Engineer, NASA Jet Propulsion Laboratory, Pasadena, CA

PROGRAM ADVISORS
Scott K. Baker, PhD, Pacific Institutes for Research, Eugene, OR
Carla C. Johnson, EdD, University of Toledo, Toledo, OH
Margit McGuire, PhD, Seattle University, Seattle, WA
Donna Ogle, EdD, National-Louis University, Chicago, IL
Betty Ansin Smallwood, PhD, Center for Applied Linguistics, Washington, DC
Gail Thompson, PhD, Claremont Graduate University, Claremont, CA
Emma Violand-Sánchez, EdD, Arlington Public Schools, Arlington, VA (retired)

TECHNOLOGY
Arleen Nakama, Project Manager
Audio CDs: Heartworks International, Inc.
CD-ROMs: Cannery Agency
ResourceLinks Website: Six Red Marbles

PHOTO CREDITS Cover © Brand X Pictures/Alamy; 1 © NASA/esa; 2-3a © Vladimir Piskunov/Shutterstock; 2-3b © Graca Victoria/Shutterstock; 2-3c © Vargyasi Levente/Shutterstock; 3a © Chuck Eckert/Alamy; 3b © coko/Shutterstock; 3c © Edward Hattersley/Alamy; 4 © Panoramic Images/Getty Images; 5, 6-7, 9a illustrations by Cam Wilson; 8 © Enigma/Alamy; 9b and 9c Lloyd Wolf for Millmark Education; 10-11 © Stocktrek Images/Getty Images; 11 © Richard J. Wainscoat; 12a, 12c, 13d © NASA/JPL-Caltech; 12b and 15b © Calvin J. Hamilton; 12d © NASA/MODIS/USGS; 12e and 15c © NASA/STScI/AURA; 13a © Elvele Images/Alamy; 13b © Reta Beebe (New Mexico State University), D. Gilmore, L. Bergeron (STScI), and NASA; 13c © Space Telescope Science Institute; 14 © David Nunuk/Photo Researchers, Inc.; 15a © NASA/Science Source/Photo Researchers, Inc.; 16-17 © Detlev van Ravenswaay/Photo Researchers, Inc.; 17 © SOHO/ESA/NASA/Photo Researchers, Inc.; 18 © ImageState/Alamy; 20 © AP Photo/Ric Francis; 21 © RTimages; 22 © Shigemi Numazawa/Atlas Photo Bank/Photo Researchers, Inc.; 24 © Stephen Mcsweeny/Shutterstock

Copyright © 2008 Millmark Education Corporation

All rights reserved. Reproduction of the whole or any part of the contents without written permission from the publisher is prohibited. Millmark Education and ConceptLinks are registered trademarks of Millmark Education Corporation.

Published by Millmark Education Corporation
PO Box 30239
Bethesda, MD 20824

ISBN-13: 978-1-4334-0101-5

Printed in the USA

10 9 8 7 6 5 4 3 2

Our Solar System: The Sun and Its Effects